Wildflowers

A WILDFLOWER IS A PLANT NOT IN CULTIVATION.

ARROW-LEAF BALSAMROOT

Balsamorhiza sagittata

Composite Family - Sunflower Tribe 1-2.5 ft.

Sagittata describes the arrowhead shaped leaves which have silver undersides. The Indians consumed seed sprouts and roots, which tasted like balsam. Bighorn sheep devour the flower heads and leaves in the spring. Livestock will eat the entire plant.

BLUEBELL

Campanula rotundifolia

Bluebell Family 6-24 in.

This is the bluebell-of-Scotland. It is sometimes called bellflower, witches-thimble, and harebell. Harebell may refer to an association with witches. According to legend, witches were able to turn themselves into hares, omens of bad luck.

Wildflowers & Weeds
OF THE PACIFIC NORTHWEST

PORTLAND GARDEN CLUB
PORTLAND, OREGON

1132 S.W. Vista Ave.
Portland, OR 97205

Library of Congress Catalog Card Number
93-86986
ISBN 0-9639462-0-X

Printed by Premier Press
Portland, Oregon

PREFACE

This book is an introduction to the study of plants. It is intended as a field guide to teach ways of identifying wildflowers and weeds of the Pacific Northwest.

We included each plant's common name and its botanical name in two words: first, the genus and second, the species.

Our charts include plant names, flower descriptions, leaves, color, habitats and blooming periods.

The book notes the use of the plants by the Native Americans, early settlers and today's people.

We hope this book will stimulate an interest in botany. Enjoy!

A WEED IS A PLANT OUT OF PLACE. A WILD FLOWER IS A PLANT NOT IN CULTIVATION.

CONTENTS

Wildflowers & Weeds

OF THE PACIFIC NORTHWEST

DEDICATED TO
SUE COOLEY

COMMON CAMAS

Camassia quamash
Lily Family 1-2 ft.

The camas bulb was a source of food for the early settlers, the Indians, and the Lewis and Clark expedition. In 1876, the Nez Perce Indians left their homes in search of better camas fields, a contributing factor in the start of the Chief Joseph Wars. Large camas fields can still be found in eastern Oregon.

CLARKIA

Clarkia amoena

Evening Primrose Family 6-24 in.

This delightful flower has many common names such as farewell-to-spring and summer's darling and herald- of-spring. Amoena means charming. The plant is named for William Clark of the famous expedition in the early 1800's. On their return trip east, Lewis and Clark found Clarkia in Idaho along the Clearwater River.

RED COLUMBINE

Aquilegia formosa
Buttercup Family 1-3 ft.

Aquilegia means eagle. The arched spurs are reminiscent of the eagle's talons. The old-fashioned columbine grown in English gardens was called granny's bonnet. The flower was chosen in Shakespeare's *Hamlet* because it was thought to be the emblem of forsaken lovers. It was part of Ophelia's bouquet.

DELPHINIUM

Delphinium nutallii

Buttercup Family 1-2 ft.

The ancient Greeks named this flower delphinium because of its resemblance to the dolphin. It was used at the Battle of Waterloo to rid the British troops of parasites. Ranchers find this plant to be toxic to their cattle. This species is listed as a rare plant in the state of Washington.

FIREWEED

Epilobium angustifolium

Evening Primrose Family 3-10 ft.

Fireweed is a pioneer plant, among the first to take over after a fire or a clearing. The common name is very descriptive, as the bright purple-pink flowers give a burned area a flame-like appearance. Northwest Indians ate the fireweed sprouts cooked or raw, and used the leaves to make a tea which is rich in vitamin C.

EXPLORER'S GENTIAN

Gentiana calycosa
Gentian Family 8-20 in.

The gentiana group is named for King Gentius of Illyria, the country Shakespeare chose for *Twelfth Night*, and which later became Yugoslavia and Albania. This king discovered the medicinal properties of the plant.

GLACIER LILY

Erythronium grandiflorum

Lily Family 6-14 in.

The glacier lily often sends its flowers up through the receding snow. Even if the flower is gone, the elliptical leaves make it easily identifiable. The Indians used the dried corms as a food source.

GRASS WIDOW

Sisyrinchium douglasii

Iris Family 6-12 in.

Botanist David Douglas found this plant at Celilo Falls in 1826. Various species of sisyrinchium are found throughout the United States. One of the treats of very early spring, it often appears on rocky slopes.

IRIS

Iris tenax

Iris Family 8-16 in.

Tenax is the word for strong or tough. The Indians used fibers from the leaf margins to make strong, pliable cords for snares and nets. Iris, though prevalent, is not a food source for livestock because of its bitter taste.

MARIPOSA LILY

Calochortus macrocarpus

Lily Family 1-2 ft.

The Spanish word mariposa means butterfly. These flowers growing in profusion appear to have many butterflies on each stem. It blooms in early summer. The genus calochortus is widely distributed and has many species, one of which, the sego lily, is Utah's state flower. All species have edible bulbs.

LEAFY PEA

Lathyrus polyphyllus

Pea Family 1-3 ft.

This is also called the Oregon pea. It is often confused with vetches. The Nootka Indians wove the beautiful flowers into their grass baskets. The sturdy and extensive root system helps prevent soil erosion.

SHOWY PHLOX

Phlox speciosa
Phlox Family 4-16 in.

The 300 species phlox family is centered in North America, the majority in the west. This species is found east of the Cascades. It has a sweet fragrance which perfumes the air around it.

CALIFORNIA POPPY

Eschscholzia californica

Poppy Family 6-16 in.

Poppies growing on the hillsides known as "lands-of-fire" were used by early mariners as markers. For its medicinal value the Indians rubbed a portion of the root on an aching tooth. The juices of the flowers and leaves cause drowsiness and relief from pain.

POET'S SHOOTING STAR

Dodecatheon poeticum

Primrose Family 6-14 in.

These flowers appear to be turned inside-out, resembling shooting stars. They grow in open, grassy areas. They are found in many areas and are easily spotted at the east end of the Columbia River Gorge.

WOODS STRAWBERRY

Fragaria vesca

Rose Family 3-7 in.

The name strawberry is derived from the practice of laying straw around the cultivated plants to keep the rain from destroying the crop. The Indians used the plants for a tea-like drink. The Quillayute Indians chewed the leaves and placed them on burns. The Skokomish Indians boiled the whole plant to use for digestive problems.

TIGER LILY

Lilium columbianum

Lily Family 2-5 ft.

The derivation of the word lily is obscure in both Latin and Greek. An English folk story stated that anyone who smelled the tiger lily would surely develop freckles. Picking the blossoms with their leaves causes the plant to starve.

TRILLIUM

Trillium ovatum

Lily Family 6-16 in.

The name trillium is derived from the fact that all parts of the plant are in threes. The leaves which nurture the plant grow so close to the bloom that, if picked, the plants usually don't have enough nourishment to produce flowers the next year.

STREAM VIOLET

Viola glabella

Violet Family 4-6 in.

These are also called johnny jump-ups or pioneer violets. They bloom profusely in the spring. In the stomach of the Tollund Man, who was buried more than 2,000 years ago in a peat bog in Denmark, a violet was found among the plants that were identifiable.

OREGON WOOD SORREL

Oxalis oregana

Wood Sorrel Family 4-6 in.

It is one of the few wildflowers found in dense redwood forests. The clover-like leaves are very sensitive and droop when there are temperature extremes. A tart juice is characteristic of the genus. The generic name is from the Greek *oxys* meaning sour.

YELLOW BELLS

Fritillaria pudica

Lily Family 5-10 in.

The blossoms turn red as they age, a characteristic which gives the species its name *pudica* meaning bashful or modest. The bulbs of this species can be eaten cooked or raw; cooked they taste like rice and raw like potatoes. They can be successfully dried for storage.

WEEDS

HIMALAYAN BLACKBERRY

Rubus discolor

Rose Family 3-8 ft.

This blackberry was introduced from the Old World. It is very invasive, found along roadsides and abandoned fields and gardens. The delicious berries are a redeeming feature, however the plant's short thorns can be hazardous.

COMMON CHICKWEED

Stellaria media

Pink Family 4-10 in.

This plant is native to Europe. It has small star-like flowers. The common name is derived from the fact that wild and domesticated birds feed on the seeds and plants. It is a menace to anyone desiring a fine lawn.

ENGLISH DAISY

Bellis perennis

Composite Family - Aster Tribe 8-20 in.

Daisy is a corruption of "day's eye" as the ray flower looks like the sun. It has an unpleasant acrid taste that makes it unappealing to cattle, but it has been used as a pot herb in times of food scarcity.

DANDELION

Taraxacum officinale

Composite Family - Chicory Tribe 6-15 in.

Dandelion is derived from the French *dent de lion* (lion's tooth). Native to Greece, it reached America shortly after the first colonists. It was used for medicine, food, and the roasted roots for a coffee substitute. Wine is made from the blossoms.

MORNING GLORY

Convolvulus arvensis

Morning Glory Family (vine)

Bindweed is very invasive and considered a pest by farmers. The rhizomes resist removal and any small portion that remains in the soil sends up a sprout. The tough network of roots, however, reduces erosion.

MULLEIN

Verbascum thapsus

Figwort Family 2-6 ft.

In imperial Rome, the mullein's stripped stalks were dipped in tallow to use as torches in funeral processions. These torches were also used by miners of the early west and were called "miners' candles." Early Roman ladies used the roots to make hair dye.

POISON OAK

Rhus diversiloba

Sumac Family 1-8 ft.

It is a noxious plant and causes severe rash ("leaves of three, let me be"). Even in winter, tearing or breaking stems can cause skin irritations. Indians used poison oak to counteract rattlesnake poisoning. Birds eat the fruit with no ill effect.

QUEEN ANNE'S LACE

Daucus carota

Parsley Family 1-3 ft.

Queen Anne's lace is the ancestor of cultivated carrots. England's Queen Anne, of the 18th century, wore this flower instead of cloth lace. According to legend the tiny red flower in the center of each cluster represented a drop of blood from Queen Anne's pricked finger.

ST. JOHN'S WORT

Hypericum perforatum

St. John's Wort Family 1-2 ft.

In ancient times, St. John's wort was thought to have several amazing properties, including the ability to foretell marriage for maidens if picked on June 24, St. John's Day. Dew gathered from its flowers was said to be an elixir for sore eyes.

SUNFLOWER

Helianthus annuus

Composite Family - Sunflower Family 2-6 ft.

Helios and *anthus* mean sun and flower in Greek. From pre-columbian times the seeds have been used for food. The seeds themselves are eaten, oil is extracted from the seeds, or the seeds are ground for flour. The flower heads follow the sun from rise to set. It is the state flower of Kansas.

TOAD-FLAX

Linaria vulgaris

Figwort Family 6-24 in.

One explanation for the name toad-flax is that if you squeeze a blossom gently from the sides, it will open like a toad's mouth. It is also called butter-and-eggs, a name derived from the flower's colors. Although it grows in great profusion, it is foul smelling and cattle avoid it.

HAIRY VETCH

Vicia villosa

Pea Family 1-3 ft.

Vicia comes from the Latin, meaning to bind together and refers to the tendrils which are typical of all 35 species of vetch. The American Indians enjoyed eating the stems and young seeds. Some species of vetch have been introduced for feeding livestock. Its roots contain bacteria which are an important source of nitrogen in the soil.

STUDYING FLOWERS

All the wildflowers shown in this book are native plants that were growing here before the white man came. All of the weeds in this book, with the exception of poison oak, are alien plants that were introduced from the Old World, mostly from Europe.

A typical flower has several parts. The essential parts are the pistils and stamens. Pistils produce ovules which, after pollination, produce seeds needed for reproduction. The stamens, resembling tiny yellow wands, produce pollen grains. The colorful petals surrounding the pistil and stamens support this activity by attracting pollinating insects. Visiting insects, attracted to particular colors and scents, carry pollen grains

SPEAK TO THE EARTH AND IT SHALL TEACH THEE. JOB 12: 8

from one plant and, in passing, deposit them on the stigma of another plant. Surrounding the ring of petals are the sepals, sometimes green and sometimes colorful to match the petals.

All of the flower parts are important to look at if you wish to identify different kinds of flowers. Look carefully at the color, shape, number and arrangement. Also, note how they are grouped. Do they grow alone at the end of a stem? Are they clustered in groups? Do they droop or stand up straight?

To identify a plant, use the illustrations on the following pages. Note the type of flower and its color. How many petals and sepals does it have? Are the flowers single, in groups, or in a tight cluster? It is best to try to identify a plant when it is flowering, but in some cases you will have to make intelligent guesses. Look at the leaves. Are the edges smooth or toothed or lobed? Are they arranged oppositely, alternately, or in whorls (circles)? Do they arise from the base of the plant? Are they single or in groups of leaflets arranged palmately or pinnately? What is their shape?

TO OUR READERS

Remember that color can vary in a plant, especially those with blue, red, or purple flowers. Yellow is usually constant.

Soil conditions, temperature, moisture and light also cause variations. The blooming time can change as well.

Common names differ from area to area. Scientific names are the same worldwide.

A WORD OF CAUTION

This book is not a guide to edible plants. Do not take a chance because there are poisonous plants and look-alikes. Do not eat any plant growing in the wild.

NAME	FLOWER	BLOOMING PERIOD
Balsamroot	A bright yellow sunflower 4" across.	April - July
Bluebell	A light blue, bell-shaped, nodding flower.	May - Oct.
Camas	Pale blue to purple flower with stamens projecting outward.	April - June
Clarkia	Pink to Rose cup-shaped flower with a crimson spot on each petal.	June - July
Columbine	A nodding bloom of 5 red sepals and bright yellow petals.	May - July
Delphinium	A flower with blue petals and blue sepals, and a hollow spur.	June - July
Fireweed	Magenta or lavender blooms 3/4"-1" wide along a straight stalk.	June - Sept.
Explorer's Gentian	5 blue petals formed in a vase-like shape and dotted with flecks of green.	August - Sept.
Glacier Lily	1-3 yellow flowers on a single stem with curving sepals.	April - June
Grass Widow	6 purple petals with yellow stamens in the middle.	March - April
Iris Oregon Flag	3"-4" wide lilac or purple flower with 3 petal-like sepals out and down and 3 erect petals curving upward.	June - July

LEAVES	HABITAT
Large, grayish leaves, arrowhead-shaped, 6" wide and 12" long.	Open fields and hillsides on both sides of the Cascades and the Columbia Gorge.
Narrow stem leaves	Rocky openings in the Cascades and Columbia Gorge.
3 or 4 grass-like leaves.	Moist meadows and swales on both sides of the Cascades.
1" - 2 1/2" lance-like leaves.	Grassy, dry slopes, open rocky & lightly wooded areas on the west side of the Cascades.
Lobed leaves split into 3 leaflets, forming a deep scalloped oval.	Moist to dry sites, wood openings at all elevations.
Leaves, rounded and indented, grow smaller as they go upward on the stem.	Partly shaded or open moist grassy slopes and meadows.
Lance-shaped leaves, alternately placed on the stem, minutely toothed & 2"- 8" long.	Roadsides, clearings and burned over areas, mostly west of the Cascades.
1/4"-3/4" long, oval leaves in opposite arrangement on the stem.	Moist meadows and rocky hillsides in the Cascades.
Elliptical leaves which are glossy and pointed.	Sub-alpine moist areas above 3,000' in the Cascades; low elevations in the eastern Columbia Gorge.
Slender, iris-like leaves from the base.	Open grassy areas, especially in the eastern Columbia Gorge.
Narrow fibrous basal leaves, to 20" long.	Forest openings, open prairies and hillsides west of the Cascades.

NAME	FLOWER	BLOOMING PERIOD
Mariposa Lily	Lavender flower with 3 green-striped petals. The 3 sepals extend beyond the petals.	July - August
Leafy Pea	Purplish-red flower on a one-sided raceme 5-13 flowers.	May - June
Showy Phlox	1/2"-1" wide pink flowers with notched petals.	April - May
California Poppy	4 petal flowers, bowl shaped and yellow to orange in color.	March - October
Shooting Star	Magenta flowers with petals bent backwards yellow cone in center.	March - May
Woods Strawberry	White flowers about 1/2" wide with a yellow center.	April - June
Tiger Lily	A yellow to orange flower nodding, with maroon spotting.	June - July
Trillium	Open flower with pure white petals.	March - May
Stream Violet	A deep yellow, 5 petaled flower with fine brown lines radiating from the center.	March - July
Orange Wood Somel	Small pink flowers, shaped like a funnel.	April - June
Yellow Bells	Yellow, bell-shaped blossoms that nod.	March - June

LEAVES	HABITAT
Long grass-like leaves.	East of the Cascades; areas that are lightly wooded and dry.
A pinnate leaf with 10-16 oval leaflets.	Open coniferous woods and forest clearings.
Opposite leaves shaped like lances.	Open woods and grasslands east of the Cascades.
Finely divided leaves.	Open areas at low elevations and roadsides.
Lance-shaped basal leaves.	Open grassy slopes and streams.
Coarsely toothed leaves divided into 3 leaflets.	Forest clearings.
Lance-like leaves are in whorls of up to 9 leaves.	Coniferous woods, especially in forest clearings.
Oval, pointed leaves in a whorl of 3.	Moist woods.
Heart-shaped leaves on an upright stem.	Moist wooded areas, especially along streams.
Leaves like clover with a notch at the tip of each leaflet.	Damp woods.
Linear leaves that are 2"-12" in length.	Open grassy areas.

NAME	FLOWER	BLOOMING PERIOD
Himalayan Blackberry	Edible berries preceded by white flowers.	June - Sept.
Chickweed	Five white petals, each petal deeply indented.	Most of the year if a mild winter.
English Daisy	Small aster white flowers with yellow centers.	June - Sept.
Dandelion	Yellow ray flowers.	May - Late Sept.
Morning Glory	Funnel-like white flowers.	May - Sept.
Mullein	Yellow Flowers on a dense spike.	June - Sept.
Poison Oak	"Leaflets three, let it be" Stay away, it is poisonous to the touch.	Late May - Sept.
Queen Anne's Lace	Dull white flowers in clusters.	June - Sept.
St. Johnswort	Yellow flowers with five petals, black dots on margins of petals.	June - Sept.
Sunflower	Flower heads 3"-5" wide central maroon disk surrounded by yellow rays.	July - Sept.
Toad-flax	Yellow elongated flowers with orange throats on a single stock.	June - Sept.
Hairy Vetch	Small purple-blue flower all on one side of the stem.	June - August

LEAVES	HABITAT
Stems with thorns leaves divided into 5 leaflets.	Pastures, roadsides and disturbed areas.
Opposite leaves oblong in shape.	Garden, lawns, waste places, and meadows.
Spoon shaped leaves.	Lawns, fields, pastures, roadsides.
Distinctive pinnately lobed leaves.	Lawns, fields, pastures, waste places.
Vine leaves arrow-shaped, climbing to 10 feet or spreading on the ground.	Gardens, meadows, roadsides, disturbed areas.
Large, gray felt-like leaves with downy felt-like hairs.	Field, roadsides, waste lands, disturbed areas.
Three shiny leaflets turning red in Fall.	Forest openings, banks, under trees, waste places.
Basal leaves finely divided with a carrot-like odor.	Fields, waste lands, roadsides, meadows.
Small opposite, simple leaves.	Fields, waste lands, roadsides, meadows.
Rough heart-shaped leaves 3"-12"	Fields, waste lands, roadsides, meadows.
Alternate simple leaves.	Fields, waste lands, roadsides, meadows.
Compound leaves with pinnately arranged leaflets, tendrils at the end.	Fields, waste lands, roadsides, meadows.

FLOWER TYPES

REGULAR

IRREGULAR

BELL-LIKE

COMPOSITE

TUBELIKE

UMBEL

SPURRED

PEALIKE

FUNNEL-SHAPED

ARUM FAMILY

SPIKE

HEAD

PARTS OF A FLOWER

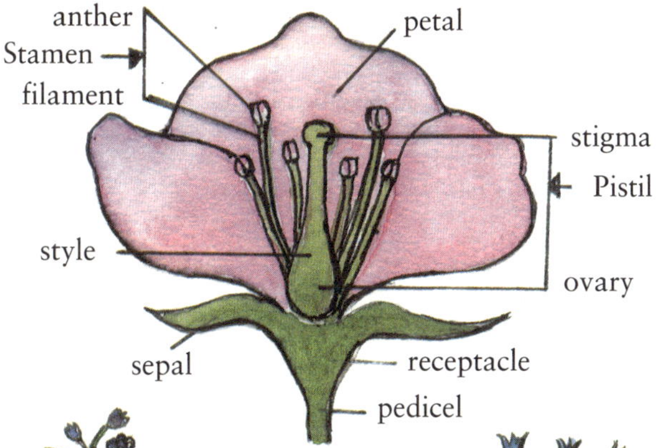

RACEME SPIKE PANICLE

FLOWER ARRANGEMENTS
ON STEM - INFLORESCENCES

LEAF STRUCTURES AND ARRANGEMENTS

HABITAT

ALPINE WOODLANDS

HABITAT

WETLANDS

GLOSSARY

Alternate - not opposite each other.

Annual - plant which completes its life cycle in one season.

Anther - the portion of the stamen where the pollen is contained.

Axil - the upper angle formed between the leaf stem and a branch.

Berry - a fleshy fruit containing seeds.

Biennial - a plant which completes its life cycle in two years.

Blade - the expanded flat portion of a leaf.

Bract - leaf-like scales on a flower cluster.

Bud - an undeveloped flower or shoot.

Bulb - a plant bud usually below the ground.

Calyx - outermost portion of a flower comprised of sepals, usually green.

Composite - flower heads composed of ray and disk flowers.

Corolla - the petals of a flower comprise this.

Corm - a bulb-like underground swelling of a stem.

Cyme - a flower cluster that blooms from the center towards the edges, the main axis is always terminated by a flower.

Disk flower - tubular florets in the center part of a composite flower head.

Divided - pertaining to a leaf cut into segments, divisions extending to the midrib or the base.

Evergreen - bearing green leaves throughout the year.

Family - a unit of biological classification.

Filament - the stalk of the stamen.

Floret - a small flower that is part of a cluster.

Flower - part of a plant containing reproductive organs.

Flower head - a dense cluster of flowers on top of a stem.

Fruit - part of a plant which bears seeds.

Genus - a biological classification below family and above species.

Habitat - the community where a plant naturally grows.

Head - a dense cluster of flowers.

Herb - a plant without woody and persistant stems above ground.

Inflorescence - the flower clusters of a plant.

Introduced - not native

Involucre - a structure beneath an inflorescence, consisting of several bracts.

Irregular - nonsymmetrical in shape.

Lobed - a leaf divided not quite halfway to the midrib.

Midrib - the vein in the center of the leaf.

GLOSSARY

Native - endemic, grows naturally in a particular place.

Naturalized - transplanted from another place, but thriving in its present location.

Nectar - sweet liquid produced by flowers to attract insects.

Nectary - a gland which secretes nectar, found on the corolla.

Node - a place on the stem where leaves or roots emanate.

Ovary - the part of the pistil that contains the developing seeds.

Palmate - having leaflets or lobes diverging from one point.

Panicle - compound flower clusters of elongated type.

Parasitic - growing on and receiving nourishment from another plant.

Pathfinders - lines that guide insects toward the nectar.

Pedicel - the stem that supports one flower.

Perennial - a plant that lives for several years.

Petiole - a leaf-supporting stem.

Pinnate - a feather-like arrangement of leaflets on both sides of a central stalk.

Pistil - the female (seed-bearing) organ of a flower.

Pod - a dry fruit that contains many seeds.

Pollen - powder-like structures produced by the stamens.

Raceme - elongated flower cluster along a single stalk.

Ray flower - flowers that encircle the disk of a composite (daisy-like) flower head, each flower may look like a petal.

Rhizome - a stem or rootstock beneath the ground.

Seedpod - sack enclosing ripened seeds.

Sepal - one segment of the calyx, usually green.

Species - the fundamental unit of plant classification, ranking below genus.

Spike - a raceme in which the flowers have no stalks.

Spine - a sharp projection from a plant.

Shrub - a woody plant near the ground with many branching stalks.

Spadix - a fleshy spike that bears flowers as in skunk cabbage.

Spur - hollow appendage of a sepal or petal.

Stamen - the organ of a flower which produces pollen.

Stigma - the end of the pistil that collects pollen.

Style - the slender stalk of a pistil.

Succulent - a thick fleshy-leafed plant that conserves moisture.

Tendril - an extension of the stem that is thin and twining.

Tuber - an underground stem that is very thick.

Umbel - A flower cluster in which all pedicels arise from a common point on a stem.

NAME THESE WILDFLOWERS

1. ____________________

2. ____________________

3. ____________________

4. ____________________

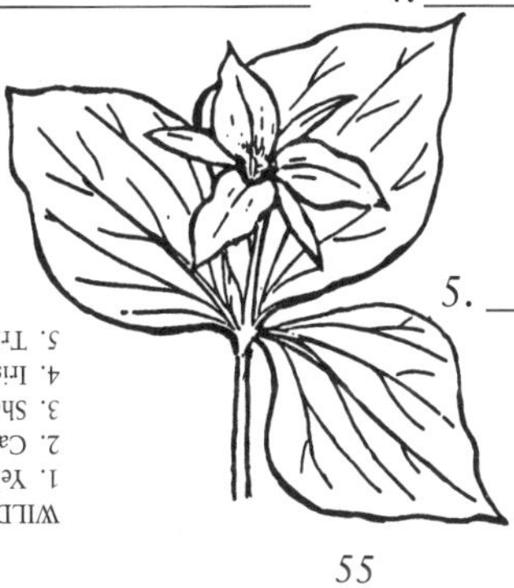

5. ____________________

WILDFLOWERS
1. Yellow Bells
2. Camas
3. Shooting Star
4. Iris
5. Trillium

NAME THESE WEEDS

1. ____________________

2. ____________________

3. ____________________

4. ____________________

5. ____________________

WEEDS
1. Morning Glory
2. Blackberry
3. Queen Anne's Lace
4. Poison Oak
5. Dandelion

SELECTED BIBLIOGRAPHY

Beals, K.M. *Flower Lore and Legend.* New York: Henry Holt and Company, 1917.

Clark, L.G. *Wild Flowers of British Columbia.* Sidney, B.C. Canada: Gray's Publishing Limited, 1973.

Craighead, J.J., F.C. Craighead Jr., and R.J. Davis. *A Field Guide to Rocky Mountain Wildflowers.* Cambridge, Massachusetts: The Riverside Press, 1963.

Crockett, J.W., and O.E. Allen. *Wildflower Gardening.* Alexandra, Virgina: Time - Life Books, 1977.

Hitchock, C.L., A. Cronquist, M. Ownbey, and J.W. Thompson. *Vascular Plants of the Pacific Northwest.* 5 Volumes, Seattle, University of Washington Press, 1969

Hollingsworth, B. *Flower Chronicles,* New Brunswick, New Jersey: Rutgers University Press, 1958.

Jolley, Russ. *Wildflowers of the Columbia Gorge.* Portland, Oregon: Oregon Historical Society Press, 1988.

Kerr, J., *Shakespeare's Flowers.* New York: Thomas Y. Cowell Co., 1969.

Kieran, J. *An Introduction to Wildflowers.* Garden City, New York: Doubleday and Co., Inc., 1952.

Lemmon and Johnson. *Wildflowers of North America.* Garden City, New York: Hanover House, 1961.

Magley, B. *Oregon Wildflowers.* Billings and Helena Montana: Falcon Press Publishing Company, 1992.

Orr, R. and Margaret Orr. *Wildflowers of Western America.* New York: Knopf Press, 1974.

Readers Digest Association Inc., *The Magic and Medicine of Plants.* Pleasantville, New York: Montreal, 1986.

Ross, R. and H. Chambers. *Wildflowers of the West Cascades.* Portland, Oregon: Timberline Press, 1988.

Spellenberg, R. *The Audubon Society Pocket Guides.* New York: Alfred A. Knopf, 1991.

Taylor, R.J. *Northwest Weeds.* Missoula, Montana: Mountain Press Publishing Company, 1990.

Underhill, J.E. *Roadside Wildflowers of the Northwest.* Blaine, Washington: Hancock House Publishers, Inc., 1981.

Whittlesey, R. *Familiar Friends, Northwest Plants.* Portland, Oregon: Roe Press, 1985.

Winegar, D. *Desert Wildflowers.* Beaverton, Oregon: Beautiful America Publishing Company, 1982.

Zim, H. and A. Martin. *Flowers,* New York: Golden Press, 1987.

CONTRIBUTORS

This work was made possible by members of the Portland Garden Club, the Katherine Pamplin Educational Endowment Fund, and Horticulture Group XI.

Concept and Illustration - Martha Pedersen

Coordinator - Carolyn Cosart

Text - Mary Ericksen and Maxine Sims

Calligrapher - Janet Charlton

Botanical Consultants - Russ Jolley, Linda McMahon, and Nancy Russell

Gwyneth Booth
Elizabeth Brooke
Susan Cooley
Don Cosart
Margueritte Drake
Elaine Flowerree
Anne Grimwood
Joan Hering
Mary Alice Hutchins
Elizabeth Kerr
Sara Mauritz
Sherman Pedersen
Richard Ritz
Patricia Schleuning
Ruby Short
Sister Patricia Stebinger

CONSERVATION AND PRESERVATION

The increased pressure of modern society has taken its toll on the natural environment. Nature's renewal of the environment progresses slowly. Consequently, we should have a heightened awareness of the need for parks and preserves where this renewal can take place undisturbed. We hope to encourage an interest in protecting the environment in which the plants in our book grow.

NOTES

OBJECT OF THE PORTLAND GARDEN CLUB

"To stimulate the knowledge and love of gardening, to aid in the protection of native trees, plants and birds, to encourage civic beauty, and to improve and protect the quality of the environment through programs and action in the fields of conservation and education."

INDEX

MOUNTAIN MEADOW

"Take only pictures, leave only footprints."

TO-DAY HE PUTS FORTH

THE TENDER LEAVES OF HOPE

WILLIAM SHAKESPEARE
KING HENRY VIII

…a wild plant,
born of a wet night,
born of an hour of sunshine;
sprung from wild seed,
blown along the roads by a
wild wind. A wild plant
that, when it blooms by chance
within the hedge of our gardens,
we call a flower; and when
it blooms outside we call
a weed;

John Galsworthy
The Man of Property.